P9-DFZ-565

THE TRIAL OF JOHN T. SCOPES

A Primary Source Account

Steven P. Olson

rosen central
Primary Source™

The Rosen Publishing Group, Inc., New York

To Dirk

Published in 2004 by The Rosen Publishing Group, Inc.
29 East 21st Street, New York, NY 10010

First Edition

Library of Congress Cataloging-in-Publication Data
Olson, Steven P.
The trial of John T. Scopes: A primary source account/by Steven P. Olson.
   p. cm.—(Great trials of the twentieth century)
Includes bibliographical references and index.
ISBN 0-8239-3974-X (library binding)
[1. Scopes, John Thomas—Trials, litigation, etc. 2. Evolution—Study and teaching—Law and legislation—Tennessee.]
I. Title. II. Series.
KF224.S3 O449 2003
345.73'0287—dc21

2002153354

Manufactured in the United States of America

Unless otherwise attributed, all quotes in this book are excerpted from court transcripts.

CONTENTS

This 1925 photograph shows John T. Scopes in the year of his famous Monkey Trial. Only twenty-four when he agreed to stand trial, Scopes (1900–1970) found his life changed forever. In his 1967 autobiography, *Center of the Storm*, he wrote, "A man's fate, shaped by heredity and environment and an occasional accident, is often stranger than anything the imagination may produce."

INTRODUCTION

At first, John T. Scopes did not want to do it. At the table in front of him were some of the leading men of the town of Dayton, Tennessee. Fred "Doc" Robinson, owner of the drugstore where they were gathered, had given Scopes a cold soda. In attendance were two brothers, Herbert and Sue Hicks, who were local attorneys, and Wallace Haggard, also an attorney. There was George Rappelyea, an excitable engineer who had sent a boy to find Scopes. Scopes had been in the middle of a tennis match, and now he was in Robinson's Drugstore, sweating in his tennis clothes, trying to decide if he wanted to risk his career for these men.

According to Arthur Blake's account of the meeting in his book *The Scopes Trial: Defending the Right to Teach*, Rappelyea had barely given Scopes a chance to sip his soda when he said, "John, we've been arguing, and I said that nobody could teach biology without teaching evolution." Scopes agreed. On a shelf at the drugstore, he found a copy of George William Hunter's *A Civic Biology*, the biology textbook that he had taught for a few weeks in April at the local high school. The book had been taught in Tennessee high schools for six years and

Inside Robinson's Drugstore in Dayton, Tennessee, some of the Scopes Monkey Trial conspirators reenact their famous meeting on July 6, 1925, for photographer George Rinhart. Left to right are George Rappelyea, Walter White, Clay Green, and Doc Robinson. After reading a press release in a local paper soliciting a teacher to stand trial, Rappelyea assembled some of the town leaders to consider bringing the case to Dayton. He was convinced that the publicity of such a case would revitalize the sleepy little town.

described in detail the theory of evolution. First proposed by Charles Darwin, the theory of evolution suggests that mankind evolved from chimpanzees over millions of years.

Then the group dropped the bomb on him. Would he be willing to stand trial for the crime of teaching evolution in the classroom?

Straight from the tennis court, Scopes had stepped into one of the most important legal contests in American history.

THE MEETING AT THE DRUGSTORE

On March 21, 1925, just six weeks before the historic meeting in Doc Robinson's drugstore, Governor Austin Peay of Tennessee signed the Butler Act, which outlawed the teaching of the theory of evolution in Tennessee schools.

Since it was first published in 1859, the theory of evolution had gathered more and more facts to support it. By 1920, most scientists supported the idea that man evolved over millions of years from lower animals. To the many Christian supporters of evolution, their religion presented a set of beliefs about morals, while the theory of evolution was a matter of scientific fact. They believed that the theory of evolution was important to education and believed that the freedom to teach it must be protected.

For others, the theory of evolution was a threat to their definition of Christianity. To these Fundamentalist Christians, every word in the Bible was true. The first book of the Bible, Genesis, states that Earth and all the living creatures on it were created by God in only six days. To Fundamentalists, the idea that man had evolved over millions of years went against the creationism in which they believed. The theory

According to Doug Linder's Web site on the Scopes trial, John Washington Butler said of the case, "The judge ought to give 'em a chance to tell what evolution is. 'Course we got 'em licked anyhow, but I believe in being fair and square and American. Besides, I'd like to know what evolution is myself." This photographic portrait of Butler taken at his desk was given to the Bryan College Archives by Butler's son.

of evolution had to be wrong. They did not want that theory to be taught in school to their children.

THE BUTLER ACT

A wealthy farmer named John Washington Butler was one such Fundamentalist believer. Butler had heard of a Tennessee girl who had returned from college believing in evolution, not in the Bible. To Butler, this news was alarming, and he soon discovered that this theory of evolution was taught in Tennessee high schools, including the one that his three teenage sons attended. In 1921, he successfully ran for state legislature on a promise to remove books that taught the theory of evolution from the classroom. Four years later, that promise became the Butler Act, which he wrote one morning after breakfast. The act prohibited the teaching of evolution in all Tennessee universities and public schools. Anyone violating the Butler Act would be fined $100 to $500.

The bill was passed in the Tennessee House of Representatives by a vote of 75–5 and then in the Tennessee Senate by a vote of 24–6. The governor, most assumed, would veto the bill and end the debate.

On March 21, 1925, Governor Peay surprised everyone by signing the bill and turning the Butler Act into Tennessee law. A deeply religious man, Governor Peay was perhaps concerned about the possible

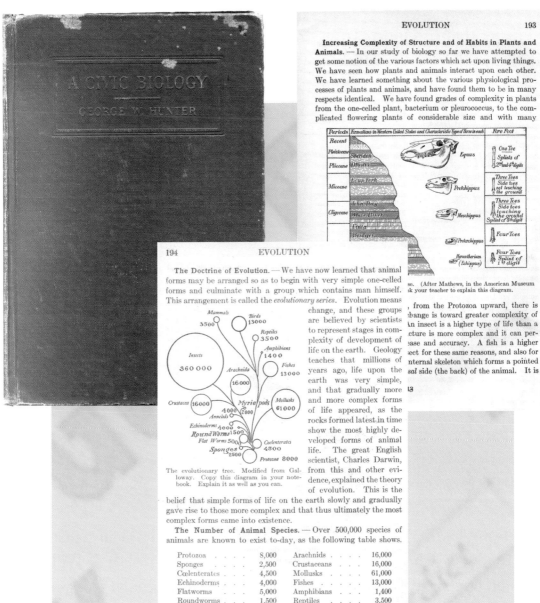

The cover and two interior pages from George William Hunter's *A Civic Biology* (1914). Like most high school biology texts of the time, the book included the theory of evolution. Two years after the Scopes trial, Hunter revised his book, removing all specific references to evolutionary theory. Perhaps in an effort to cater to Fundamentalists, he added the words, "Man is the only creature that has moral and religious instincts."

effects that teaching evolution would have on the religious views of schoolchildren. Or perhaps, he did not want to be the one to stand in the way of this law, for when Governor Peay signed it, he said in a special message to the state legislature:

> After a careful examination I can find nothing of consequence in the books now being taught in our schools with which this bill will interfere in the slightest manner. Therefore it will not put our teachers in any jeopardy. Probably the law will never be applied. It may not be sufficiently definite to admit of any specific application or enforcement. Nobody believes that it is going to be an active statute.

How wrong he was.

THE ACLU RESPONDS

A few days later, Lucille Milner, the executive secretary of the American Civil Liberties Union (ACLU) in New York, read a newspaper story on the Butler Act and brought it to the attention of the ACLU board of directors.

Founded in 1920, the ACLU seeks to provide legal assistance to individuals whose civil liberties have been violated. Whenever Americans have faced discrimination because of their race, gender, or beliefs, the ACLU is often the first organization to volunteer its legal help. The actions of the ACLU have angered many people. But to the people who need its help, the ACLU is a champion of freedom and provides protection through legal action. At the board meeting in 1925, the ACLU decided what action to take.

On May 4, 1925, a press release appeared in the *Chattanooga Daily News* declaring, "[The ACLU is] looking for a Tennessee teacher who is willing to accept our services in testing this law [the Butler Act] in the courts . . . By this test we hope to render a real service to freedom of teaching through the country, for we do not believe the law will be sustained." The ACLU believed that the Butler Act was unconstitutional, meaning that it violated the U.S. Constitution. When a law is declared unconstitutional by a state court, it is no longer a law. However, a higher court may choose to review that decision on appeal, and if the law pertains to U.S. constitutional rights, the United States Supreme Court can choose to make the final decision on the matter. The process of appeals can take years to complete, and the ACLU was prepared for a long fight.

SCOPES GIVES IN

It was the ACLU notice in the Chattanooga paper that George Rappelyea read on the morning of May 5, 1925, and brought to Robinson's Drugstore. A Christian and a believer in the theory of evolution, Rappelyea was more interested in the excitement that the case would create. If the case was tried in Dayton, it would bring people and business to the town. He had no trouble convincing Fred Robinson, who served as the chairman of the school board, and Walter White, the superintendent of the local school district, that Dayton should try the case. Also present and enthusiastic were the three attorneys: the Hicks brothers and Wallace Haggard.

But who would be the teacher willing to test the case? The Rhea County Central High School's regular biology teacher, W. F. Ferguson, had a family to support. If he chose to participate in the trial, he would probably lose his job. It was then that Rappelyea sent a local boy to find Scopes, a young and single man.

Curiously, Scopes was not a biology teacher. At Rhea County Central High School, he taught math, chemistry, and physics. He also coached several sports teams. He had substituted for the ailing Mr. Ferguson the previous month.

Taken in Dayton on July 18, 1925, this photograph shows George Rappelyea *(left)* with John Scopes. Even though he was one of the instigators of the trial, Rappelyea never claimed to be an evolutionist.

While he knew that standing up for the right to teach evolution would be a risk to his career, Scopes felt that he had to do it. He had been raised in a family that believed in academic freedom. His father had read to him about the theory of evolution when he was a child. Finally, John T. Scopes agreed to be their test teacher.

The next day, Rappelyea dashed off a telegram to the ACLU. Scopes eventually found a Dayton sheriff's deputy, Burt Wilbur, to serve him a warrant. Scopes was to appear before three justices of the peace on May 9. Thus the wheels of the American justice system began to turn to resolve the debate over teaching evolution in the classroom.

EVOLUTION VERSUS CREATION

CHAPTER
2

On December 27, 1831, Charles Darwin started work as a naturalist on an expedition of the South American coast and selected islands in the Pacific aboard a ship called the *Beagle*. Over the next five years, the scientist gathered species and recorded the observations that would become the foundation of his great work *On the Origin of Species by Means of Natural Selection, or the Preservation of Favoured Races in the Struggle for Life*.

Published in 1859, *The Origin of Species* argued that all life, including human beings, changed over time. These changes occurred in very small steps between parent and child. The child was a bit different from the parent, who was a bit different from the grandparent, and so on. Over many generations, a species could change a great deal.

Sometimes these changes benefited a given species, and sometimes they did not. Darwin observed that the struggle for life is very difficult. The species that won the struggle changed in ways that provided advantages over other species. Those species that gained advantages outlived those that did not. Darwin called this struggle "natural

ON

THE ORIGIN OF SPECIES

BY MEANS OF NATURAL SELECTION,

OR THE

PRESERVATION OF FAVOURED RACES IN THE STRUGGLE
FOR LIFE.

BY CHARLES DARWIN, M.A.,

FELLOW OF THE ROYAL, GEOLOGICAL, LINNEAN, ETC., SOCIETIES;
AUTHOR OF 'JOURNAL OF RESEARCHES DURING H. M. S. BEAGLE'S VOYAGE
ROUND THE WORLD.'

LONDON:
JOHN MURRAY, ALBEMARLE STREET.
1859.

The right of Translation is reserved.

MR. BERGH TO THE RESCUE.

THE DEFRAUDED GORILLA. "That *Man* wants to claim my Pedigree. He says he is one of my Descendants."

MR. BERGH. "Now, Mr. DARWIN, how could you insult him so?"

selection." Whether a species was the fastest in the jungle, able to blend into its surroundings, armed with a poisonous bite, or covered in sharp spines, it survived based on the development of these useful characteristics. The process of this struggle for survival over millions of years was called evolution.

When the famous biologist Thomas Huxley read the book, he wrote back to Darwin, "As for your doctrine, I am prepared to go to the stake." In the next few years, Darwin's theory gained acceptance among most scientists. In 1871, he published *The Descent of Man* and *Selection in Relation to Sex*, in which he argued that mankind descended from more primitive animals and had changed over millions of years. Like the creatures that Darwin had researched on the voyage of the *Beagle*, mankind had evolved through many small changes over many generations.

THE RISE OF FUNDAMENTALISM

From 1880 to 1920, the United States witnessed a great deal of change. Several new states joined the country. The Industrial Revolution brought many different cultures into the melting pot of America. Many of these new arrivals had different beliefs than the Americans who had arrived earlier. Values were changing, too. Women wanted equal rights, including the right to vote. Jazz music and wild dancing were sweeping the nation. Although it was prohibited by law, alcohol

Charles Darwin *(top left)*, the naturalist who published his theories of evolution in the controversial 1859 book *On the Origin of Species by Means of Natural Selection (top right)*, was born in England in 1809. In Thomas Nast's cartoon drawing at the bottom of the page, animal rights activist Henry Bergh scolds Charles Darwin for insulting a gorilla by claiming that man descended from apes. The cartoon was created in 1871.

seemed to be everywhere. For some people, the amount of change in America was very troubling.

These changes in society worked their way into Christian churches throughout America. Some ministers began to preach that the theory of evolution may have some truth to it. For churchgoers who believed every last word of the Bible, it was too much. These Christians began to leave their churches in search of a different vision.

In 1910, two executives of the Union Oil Company published the first of a series of pamphlets entitled *The Fundamentals—Testimony to the Truth*. Delivered to three hundred thousand ministers, these pamphlets established the basic ideas of the Fundamentalist movement. The pamphlets argued that Jesus, the son of God, was born to a virgin, died for our sins, and rose from the dead to rejoin God in heaven. They argued that the stories of the Bible were truly the word of God and that there was no room for interpretation. The Book of Genesis was right, and Charles Darwin was wrong.

The Fundamentals gained acceptance across the country among Christians from different churches. The movement became organized when Dr. William Bell Riley founded the World Christian Fundamentals Association in 1919. New beliefs, such as evolution, clouded the picture of the world as the Fundamentalists saw it. So, in the minds of Fundamentalists, laws that permitted these new beliefs had to be stopped.

In 1922, a bill to outlaw the teaching of the theory of evolution was narrowly defeated in the Kentucky Senate. A similar bill was debated in South Carolina in the same year. In 1923 and 1924, a number of legislatures in the South confronted the issue of teaching evolution, and small gains were made for the Fundamentalist movement.

Then, in 1925, the Fundamentalists scored their first great victory with the Butler Act in Tennessee.

THE ISSUES OF THE TRIAL

As the forces began to gather for the trial in Dayton, Tennessee, it became apparent that the prosecution and the defense would not be arguing different sides of the same question, as is customary in a trial. For the prosecution, the question was simple: Did John T. Scopes violate the law that had been passed in the Tennessee legislature? For the defense, the question before the court was whether the law was constitutional: it was clear that Scopes had violated the law by teaching evolution to his students. But did the state of Tennessee have the right to limit the freedom of speech of teachers?

Behind these different questions lay more conflicts. There were conflicting views about the role of change in the country. There were conflicting views about evolution versus creationism, about the separation of church and state, and about the government's power versus taxpayers' rights. All of these conflicts were gathering like a storm in tiny Dayton, Tennessee, and, through the magic of radio, were spreading across the country in what became known as the Trial of the Century.

THE NATION TAKES SIDES

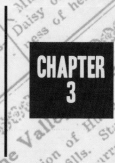

On May 9, 1925, John Scopes and George Rappelyea presented a copy of the Butler Act and a copy of the biology text that Scopes had taught in April to three Dayton justices of the peace. Scopes was ordered to appear before a grand jury in August and was released on bond.

A few days later, leaders in Chattanooga attempted to set up their own trial with their own teacher. In an effort to remain first, Scopes was called back from vacation and forced to appear before a special meeting of the grand jury on May 25. He was charged with violating the Butler Act. The trial was set for July 10 at the courthouse in Dayton.

As soon as the trial date was set, lawyers from all over the country volunteered their services. Jumping on the bandwagon for the prosecution was the finest orator of the time, William Jennings Bryan.

WILLIAM JENNINGS BRYAN, THE GREAT COMMONER

By 1925, William Jennings Bryan was sixty-five and in the twilight of a great career. As a young attorney in Nebraska, Bryan ran for the U.S.

House of Representatives with a promise to protect the interests of people without power. He won. Thus, the Great Commoner was born.

Bryan's popularity grew during his two terms in the House. Bryan supported many of the issues that are taken for granted today. He believed in a woman's right to vote, election of senators by popular vote, and the formation of a Department of Labor.

At the Democratic Convention in 1896, he gave his famous "Cross of Gold" speech before the delegates and won the Democratic nomination for president. Although Bryan did not win the presidency in any of the years he was nominated (1896, 1900, and 1908), he did bring to national attention the causes of poor working people and farmers. By most accounts, he was a just and decent man.

This 1925 photograph shows the famous politician and orator William Jennings Bryan preaching at a Sunday service at the Little Methodist Church in South Dayton, Tennessee. Bryan's Fundamentalist beliefs led him to take the case against John Scopes, and they also gained him favor among the locals.

The root of Bryan's pursuit of justice lay in his deep beliefs in the Bible. In his latter years, Bryan took up the cause of Fundamentalism and began to use his speaking skills to spread those beliefs around the country.

Although Bryan had not practiced law in thirty years, several experienced trial attorneys joined him and three local attorneys on the prosecution team. A. Thomas Stewart was the attorney general of the Eighteenth Judicial Court. A straightforward attorney, he led the prosecution. Benjamin G. McKenzie, a former assistant attorney

Members from the defense team meet with William Jennings Bryan and some of his prosecution team before the start of the Scopes trial on June 23, 1925. Left to right sit J. Gordon McKenzie, William Jennings Bryan, Wallace Haggard, Herbert Hicks, and Harry Lawrence.

general, and his son J. Gordon McKenzie were well-known locals from Dayton. Bryan's son, William Jennings Bryan Jr., also participated.

THE DEFENSE

When William Jennings Bryan volunteered his services, the Hicks brothers fired off a letter to him, writing, "We will consider it a great honor to have you with us in this prosecution." The defense team knew the stakes in the trial had been raised and that the trial would get more publicity from the participation of a famous attorney.

At that point, the defense team consisted of only a few attorneys. John Neal was a former dean of the University of Tennessee law school and a scholar of the U.S. Constitution. He had run against Austin Peay for governor and had lost by a wide margin, in part because of his opposition to the anti-evolution movement. Assisting him was his associate, F. B. McElwee. From New York came Dudley Field Malone, a fine attorney and speaker, and Arthur Garfield Hays, a lawyer provided by the ACLU, who was responsible for planning and preparing an appeal.

The defense team did not think it could win the case, so it sought to get the case appealed to a higher court, where the legality of the Butler Act could be judged. If a higher court reviewed the appeal and found errors in the judge's handling of the case, then it could be heard in that court.

Still, the defense team had no one to match Bryan's popularity. Fortunately for Scopes, Bryan and the Fundamentalists angered one of the most experienced and famous attorneys in America.

This portrait of Clarence Darrow (1857–1938), was taken in 1922. A defender of the underdog and a determined opponent of capital punishment, Darrow fought to better America. As he recognized in his opening statement, "Scopes isn't on trial, civilization is on trial."

CLARENCE DARROW, ATTORNEY FOR THE DAMNED

For two men sitting on opposite sides of a courtroom, William Jennings Bryan and Clarence Darrow had much in common. Like Bryan, Darrow was a

Surrounded by his defense attorneys, John Scopes prepares for his trial in this photograph taken in 1925. Seated from left to right are Clarence Darrow, Dudley Field Malone, John Neal, John Scopes, and Fred Robinson. George Rappelyea stands on the left with two unidentified men.

champion of the rights of the underprivileged and was either loved or hated for his work. He had risen to fame by representing Eugene V. Debs, leader of the American Railway Union, in an important trial that had earned him the reputation as Attorney for the Damned.

In the following years, Darrow defended more labor leaders, anarchists, African Americans, and others who could not find a lawyer to stand beside them. Shortly before the Scopes trial, Darrow had been in the national spotlight for defending Nathan Leopold and Richard Loeb, two friends on trial for murdering a fourteen-year-old boy. The young men had already confessed to the crime, but Darrow managed to save them from the electric chair. Darrow's folksy ways and wrinkled suits hid a fighter with a tremendous legal mind.

In one critical way, Darrow differed from Bryan. While Bryan was a leader of the Fundamentalist movement, Darrow was a committed agnostic. An agnostic does not have a final opinion on whether God exists. Yet, Darrow did believe in the legal system of the United States, including the freedom of speech. In defense of that freedom and of John T. Scopes, Darrow offered to help. As he later noted in his autobiography, *The Story of My Life* (1932), "For the first, the last, the only time in my life, I volunteered my services in a case; it was in the Scopes trial in Tennessee that I did this, because I really wanted to take part in it."

In early July in a broiling courtroom, the trial began.

THE TRIAL BEGINS

As the trial date approached, people from all over the country streamed into Dayton. Located on the banks of the Tennessee River, Dayton had been a mining town of five thousand people at the turn of the century. But the mines went dry, the coal companies went bankrupt, and by 1925, the population had dwindled to eighteen hundred. In a few days, this forgotten town turned into the most important city in America. Everyone, it seemed, was coming to Dayton.

Many people came by the extra trains sent up from Chattanooga. Some landed at the first airstrip in Rhea County, created for the event. Some came by car, and some seemed to crawl out of the hills surrounding Dayton. In his autobiography, John Scopes called them "one of the rarest collections of screwballs" he had seen in his life.

Almost overnight, downtown Dayton became a circus. Up and down Main Street, revivals and prayer meetings were led by preachers of all sorts. Fundamentalists built an open-air tabernacle. Signs suggested "Read Your Bible," "Come to Jesus," and "Prepare to Meet the Maker." Other signs were just clever advertisements. In the front window of J. R.

If his motive was to bring publicity to Dayton, George Rappelyea got his wish. Hordes of spectators descended upon the quiet southern town. Businesses used any angle they could think of to tie the court case to their wares. In the photograph above, taken on July 7, 1925, visitors crowd street vendors selling souvenirs of the trial. It seems the evolution versus creation debate had struck a chord in nearly everyone.

The Trial of John T. Scopes

Darwin's Everything to Wear Store, passersby read, "Darwin Is Right—Inside." Even Robinson's Drugstore had one: "Where It All Started." Inside, shoppers could buy a Simian Soda from Doc Robinson himself.

Out in the street, vendors sold food and souvenirs fit for a carnival. Hot dogs, lemonade, ice cream, soda, and watermelon were peddled from the backs of trucks. Tourists could buy stuffed monkeys and a variety of pins to take back home. A live chimpanzee even came to perform.

As a symbol of evolution, the monkey became Dayton's mascot for the Scopes trial. Even though the case was serious business and the debate was often tense and brutal, street hawkers managed to bring some fun to the matter. In this July 11, 1925, photograph, a girl holds a monkey doll and wears a sign bearing the slogan, "They can't make a monkey out of me."

More than two hundred reporters from major newspapers all over the country, plus Britain, France, and Germany, landed in Dayton. WGN Radio from Chicago came to make the first live radio broadcast in history. Behind the grocery store, Western Union telegraph operators would tap out two million words over the telephone lines to a world waiting to hear from Dayton, Tennessee.

THE LAWYERS ARRIVE

Three days before the start of the trial, the Royal Palm Limited from Miami arrived at the Dayton train station. William Jennings Bryan stepped from the rear platform and doffed his white pith helmet to an applauding crowd. In one of his speeches before the start of the trial, Bryan declared, "The contest between evolution and Christianity is a duel to the death."

The night before the trial started, Clarence Darrow stepped off the train to a much quieter reception. As he later recalled in his autobiography, "There was no torchlight parade to greet me." Darrow made an appearance at a dinner and then retired to rooms provided by a local family, for he was due in court the next day.

JUDGE JOHN RAULSTON, PRESIDING

On July 10, 1925, more than seven hundred people crowded into the second-floor courtroom in the Rhea County Courthouse on what promised to be a sweltering day. While many of the lawyers, witnesses, reporters, and audience members fanned themselves with palm leaves, Judge John Raulston waited for the WGN Radio team to adjust their equipment so that his gavel could be heard around the world.

Born and raised in Tennessee, Judge Raulston liked to call himself just a regular mountain judge. A short and thickset man, he brought his Bible with him into court, and it made an immediate appearance at the

This photographic portrait of Judge John D. Raulston was taken on July 11, 1925. Although Raulston was a deeply religious man, he applauded Scopes's and Darrow's courage by citing this definition of greatness: one who "possesses a passion to know the truth and [has] the courage to declare it in the face of all opposition."

trial, which opened with a long prayer. Clarence Darrow was irritated by the praying, but he chose not to object to it in the beginning. The defense had many battles yet to fight.

JURY SELECTION

That afternoon, more than one hundred men were brought into the courtroom for jury selection. At that time in Tennessee, women were not permitted to serve on juries. To each prospective juror, Clarence Darrow asked three questions: Did he know about evolution? Did he have an opinion about it? Would his mind be made up based on the evidence presented in the case? Darrow could excuse three possible jurors without presenting a reason, but any other juror had to be excused through argument. Darrow could not predict the answers from the rest of the jurors without interviewing them. In the end, eleven out of the twelve jurors were church members. As Darrow noted of the jury, "It is as we expected."

CASE DISMISSED?

Over the weekend, both sides jockeyed for position. Bryan delivered a sermon to the Methodist church with Judge Raulston in the front row and followed it with a speech on the courthouse lawn before three

Taken in 1925 in Dayton, this photograph records eleven of the twelve Scopes jurors. *Bottom row, left to right:* W. G. Taylor, J. B. Goodrich, J. R. Thompson, W. G. Day, R. L. Gentry, J. S. Wright. *Top, left to right:* Sheriff R. R. Harris (standing), R. L. West, W. D. Smith, J. W. Riley, J. W. Dagley, J. Bowman, and Judge Raulston (standing). The twelfth juror, W. R. Roberson, does not appear in the photo.

thousand people. Darrow, meanwhile, spoke to reporters from bed, as the rest of his quarters were stacked with legal papers. Both sides were ready for the fight.

On Monday, court opened with the formal reading of the indictment, which the defense moved to suppress. Attorney John Neal argued that the Butler Act violated the Tennessee Constitution on fourteen counts and that it violated the First and the Fourteenth Amendments to the

U.S. Constitution. To preserve the possibility of an appeal, Neal focused on the Butler Act's violations of Tennessee's constitution. Before the jury, he noted, "The legislature spoke for the majority of the people of Tennessee, but we represent the minority, the minority that is protected by this great provision [freedom of speech] in our constitution." The prosecution countered with an argument based on the Tennessee legislature's legal right to determine what is taught in school. Then Darrow stood and delivered a brilliant speech over two hours that concluded with a warning:

> We are marching backwards to the glorious age of the sixteenth century when bigots lighted fagots [sticks] to burn men who dared to bring any intelligence and enlightenment and culture to the mind.

That night, a powerful storm knocked out the power in Dayton. Some took it as a sign of God's displeasure with Clarence Darrow.

THE PROSECUTION

As usual, Judge Raulston opened court on Tuesday, July 14, with a prayer. This time, Clarence Darrow immediately objected: "This case is a conflict between science and religion, and no attempt should be made by means of prayer to influence the deliberation and consideration of the facts in the case." A silence fell in the room. As he noted later in his autobiography:

> The people assembled looked as though a thunderbolt had stunned them, and the wrath of the Almighty might be hurled down upon the heads of the defense. None of them had ever heard of any one objecting to any occasion being opened or closed or interspersed with prayer.

Judge Raulston was stunned, too. For a man who brought the Bible with him every day to work, there was little harm in prayer. After he overruled the objection to prayer, Judge Raulston read a six-thousand-word ruling on the defense's constitutional objections and their request to suppress the indictment against Scopes. Point by point,

A religious man living in a religious community, Judge Raulston thought nothing of leading the room in prayer at the start of each court session. Clarence Darrow fought hard to stop the practice, arguing that it would influence the jury. This news photograph, taken on July 15, 1925, shows the Dayton, Tennessee, courtroom opening a trial session with prayer. William Jennings Bryan, head bowed, stands directly behind the microphone.

Raulston stepped through each motion of the defense and overruled each one. Daily prayer continued through the rest of the trial.

OPENING STATEMENTS

On Wednesday morning, Judge Raulston asked the defense for their plea. John Neal rose and stated, "Not guilty, may it please Your Honor."

The prosecution and the defense presented their opening statements. Stewart's statement for the prosecution was short and to the point. The prosecution, he said, would prove that Scopes had broken the law of the state of Tennessee by teaching evolution in the classroom.

The defense had wider aims. While Dudley Malone claimed that John Scopes was innocent of committing a crime, the defense also intended to prove that the poorly written Butler Act actually charged Scopes with two crimes. According to the law, Scopes was charged with teaching a theory that opposed the Bible's version of creation. For the second crime, he was charged with teaching that humans had evolved from lower animals. While the defense was willing to admit that Scopes had taught the theory of evolution, Malone argued that there was no conflict between evolution and the story of the Bible:

> We shall show that there are millions of people who believe in evolution and in the stories of creation as set forth in the Bible, and who find no conflict between the two . . . We maintain that science and religion embrace two separate and distinct fields of thought and learning.

It was a smart plan. Bryan and the prosecution saw the trial as a "duel to the death." To them and many others, the choice was the Bible or evolution. Malone and the defense knew that they could not win that fight, particularly in this Tennessee courtroom. Instead, they sought to present Christianity as a guide to moral behavior and evolution as a matter of scientific investigation. If the defense could make that argument in this court or a higher one, then it might win the case eventually.

WITNESSES FOR THE PROSECUTION

After a lunch break, Judge Raulston swore in the jurors. Three and a half days after the trial began, the first witness was called.

This 1925 photograph of Dayton, Tennessee, during the Scopes Monkey Trial shows the position of most area residents. On one of the town's streets, an antievolution league set up headquarters and a book fair, from which they peddled publications that supported keeping evolution out of the schools. Signs posted at the headquarters read, "Hell and the High School."

Walter White, superintendent of schools in Rhea County, moved to the witness stand. Although White, a supporter of evolution, had been one of the first to convince Scopes to stand trial, he was called to the stand as a witness for the prosecution. As superintendent, he set teaching policy at the high school and could testify whether Scopes followed that policy. He was also at the meeting at the drugstore. White testified about the meeting: "Scopes told me that he had reviewed the entire book during certain days in April . . . and among other things he said he could not teach that book without teaching evolution." At this point, prosecutor Stewart entered the King James Bible into evidence as a reference to the Bible. Under cross-examination, White admitted to Darrow that the textbook in question, *A Civic Biology*, had been taught in Tennessee schools for six years. After the passage of the Butler Act, White had failed to tell any of his teachers, including Scopes, about the new law. Darrow presented a different picture of Scopes: a teacher simply doing his job.

Next to the witness stand came two of Scopes's students. Howard Morgan, fourteen, testified he had learned that long ago:

This news photograph, taken on July 15, 1925, shows Rhea County superintendent of schools Walter White on the stand testifying that Scopes did indeed teach the theory of evolution to his students. Although White was a supporter of evolution and one of the conspirators from Robinson's drugstore, he was called as a witness for the prosecution.

> . . . there was a little germ of a one-celled organism formed, and this organism kept evolving until it got to be a pretty good-sized animal, and then came on to be a land animal, and it kept on evolving, and from this was man.

Seventeen-year-old Harry Shelton testified that Scopes had presented the theory of evolution in class in April. During cross-examination, Darrow asked Shelton if he was a church member. "Yes, sir," he said. "You didn't leave the church when he told you all forms of life began with a single cell?"

Shelton answered, "No, sir."

Doc Robinson, owner of the drugstore and chairman of the school board, then testified that on May 5 he had heard Scopes admit to teaching evolution in the classroom.

The prosecution read into evidence the first two chapters of the Book of Genesis, which cover the story of creationism. When the prosecution finished telling the story to the court, the jury, and the packed courtroom, it closed its case.

THE DEFENSE

For the volunteer defense attorneys, the trial of John T. Scopes was more than a criminal case. The issue was not about whether Scopes was innocent or guilty or even if evolution was right or wrong. The case represented a chance to teach the nation, through massive media coverage, that teaching evolution was protected under the First Amendment: the freedom of speech.

Many people use the term "freedom of speech" incorrectly. The First Amendment suggests that Americans can say whatever they want. This is not true. Legal cases on the First Amendment have established ground rules for its use. A person cannot make a hateful speech against another. A person cannot make speeches in a classroom that are disruptive to the purpose of the class. Since Scopes was teaching science to the science class, his actions could not be taken as a violation of the First Amendment. However, the defense did have to establish that the theory of evolution was accepted science and that many scientists believed in it as well as in the biblical story of creation. That was their strategy.

NO EXPERTS IN EVOLUTION

On Thursday, July 16, the defense called its first witness to the stand: Dr. Maynard Metcalf, a former professor of zoology at Oberlin College in Ohio. Metcalf was an expert in the theory of evolution.

Since the first day of the trial, the defense had been arguing with Judge Raulston to allow expert testimony from scientists such as Dr. Metcalf. In court, an expert can testify about his or her specialty if such testimony is considered helpful or relevant by the presiding judge. Expert testimony is considered more important than that of a nonspecialist. As Darrow called Dr. Metcalf to the stand, Judge Raulston still refused to make a final ruling on the use of experts. Dr. Metcalf would be allowed to testify for the defense. Then, Raulston would rule.

One of the foremost authorities on zoology, Dr. Metcalf was also a regular member of and Bible teacher in the Congregational church where Bryan had preached over the weekend. This scientist would demonstrate that belief in evolution did not conflict with belief in the Bible.

Darrow asked Metcalf directly, "Do you know any scientific man in the world that is not an evolutionist?" Metcalf responded:

> I am acquainted with practically all of the zoologists, botanists and geologists of this country—I know there is not a single one among them that doubts the fact of evolution.

Dr. Metcalf provided a definition of evolution: the change of an organism from one character into a different character by means of changes in structure, behavior, function, or its method of development. At this point, Judge Raulston forbade reporters from publishing any of Dr. Metcalf's testimony in their newspapers. Shortly thereafter, he asked Darrow to end the questioning, "Will this extend very much further? It has been a pretty hard day for me." Darrow ended as asked,

Top: Taken in July 1925 in Dayton, Tennessee, this photograph captures many of the important participants in the Scopes trial, including scientists gathered by Darrow to testify as experts. *Front row, left to right:* Wilbur Nelson, Fay Cooper Cole, W. C. Curtis, H. H. Newman, and J. G. Lipman. *Middle row, left to right:* John Neal, Maynard Metcalf, Charles Potter, W. L. MacLaskey, W. H. Kepner, Arthur G. Hays, and J. N. Wheelock. *Back row, left to right:* E. Heldeman-Julines, George Rappelyea, Frank Flonozas, and Watson Davis.

Inset: Dr. Maynard Metcalf, an evolution expert, sits for a 1925 photographic portrait. Metcalf's testimony was intended to show the jury that belief in the theory of evolution need not contradict religious ideals.

and he left court expecting to continue his questioning of Dr. Metcalf in the morning.

BRYAN DELIVERS HIS FIRST SPEECH

As court opened on the next sweltering day, Stewart protested that the use of experts in the case was unnecessary, as their testimony had no bearing on whether John Scopes had broken the law or not. Darrow countered that these experts were important to providing a clearer definition of evolution. After Dr. Metcalf, Darrow had eight more experts to call; without them, the defense had no case. The jurors were removed from the room, and lawyers from both sides argued the issue for hours.

After lunch, William Jennings Bryan, the great orator, rose to speak for the first time. Standing in a white shirt and black bow tie, Bryan created an imposing figure. His many fans in the audience had been waiting for this moment.

"My friends," he began. He immediately corrected himself, "I have been so in the habit of speaking to an audience instead of a court that I will sometimes say 'my friends,' although I happen to know that not all of them are my friends." The audience laughed at the poke at the defense team. Bryan argued that the people of the state of Tennessee knew what they were doing in passing the Butler Act:

> It isn't proper to bring experts in here to try to defeat the purpose of the people of this state by trying to show that what they denounce and outlaw is a beautiful thing that everybody ought to believe in.

Loud applause filled the room. Bryan proceeded to attack Dr. Metcalf's testimony. He attacked Clarence Darrow's involvement in the Leopold and Loeb case. He continued to work the courtroom as if it were a religious revival. The people of Tennessee, he declared, were the real experts here:

> When the Christians of this state have tied their hands and said, "We will not take advantage of our power to teach religion to children by teachers paid by us," these people [gesturing at the defense table] come in from outside . . . and force upon the people of this state a doctrine that refuses not only their belief in God but their belief in a Savior and belief in heaven and takes from them every moral standard that the Bible gives us!

The room exploded in cheers. People leapt from their chairs, some crying, to congratulate Bryan for his fine speech. Clarence Darrow turned to Arthur Hays and asked, "Can it be possible that this trial is taking place in the twentieth century?"

From the defense table rose Dudley Field Malone. In nearly a week of spirited court battle in temperatures over 100 degrees, Malone had yet to remove his coat. Now, he did. Malone began:

> We have been told that this was not a religious question. I defy anybody, after Mr. Bryan's speech, to believe that this was not a religious question.

Since Bryan had made it a fight between religion and evolution, Malone continued, experts on evolution were required to support the defense's view that the Bible taken literally was not an authority in a court of law. Malone stared directly at Bryan and said:

> We feel we stand with progress. We feel we stand with science. We feel we stand with intelligence. We feel we stand with fundamental freedom in America. We are not afraid.

When Malone concluded, an applause even greater than the one given to Bryan rattled the room. As court emptied for the weekend,

MY FIRST REAL BATH. GEE! AINT IT GREAT!

DAYTON

PUBLICITY

Bryan remained in his seat. Recalling the moment forty years later, Scopes wrote that Bryan said, "Dudley, that was the greatest speech I have ever heard," according to Arthur Blake's *The Scopes Trial.*

Malone replied softly, "Thank you, Mr. Bryan. I am sorry it was I who had to make it."

Judge Raulston ruled that experts would not be allowed to testify. While they could read statements into the trial record, their testimony would not be heard by the jury, the reporters, or the radio audience. Darrow protested that the judge had overruled every single defense motion during the trial. Judge Raulston shot back, "I hope you do not mean to reflect upon the court."

Darrow replied, "Well, Your Honor has the right to hope." Hope, it seemed, had left the building for the defense team. Without expert witnesses, the defense had no one else to call to the stand, and it appeared Scopes had no chance. But Darrow had one last card to play.

William Jennings Bryan makes his first speech in the Scopes trial in this 1925 photograph. In the political cartoon beneath it, Bryan represents the sun shining down on the town of Dayton, seen as a child bathing in a pool of publicity. The drawing, titled "My first real bath: gee! Ain't it great!" was created by the artist Clifford Kennedy Berryman (1869–1949) and was probably published in the *Washington Star*. It illustrates Dayton's desire for fame overshadowing the issues of the trial.

DARROW VERSUS BRYAN

Over the weekend, the town of Dayton emptied. The carnival in the street packed its trucks and wagons and disappeared into the Tennessee hills. Most reporters piled back onto the train, figuring that the trial was finished. Those who left missed what the *New York Times* later called "the greatest court scene in Anglo-Saxon history."

On Monday morning, Judge Raulston formally charged Darrow with contempt of court for his remarks on Friday. When Darrow apologized to the judge in court, the charge of contempt was dropped. In his response, the judge said, "The Man that I believe came into the world to save man from sin . . . taught that it was godly to forgive . . . I believe in that Christ. I believe in these principles. I accept Colonel Darrow's apology."

TO THE FRONT PORCH

After lunch, Judge Raulston moved the trial to the front porch of the courthouse because of cracks in the first-floor ceiling. It was assumed that the defense would soon close its case. Darrow and the other defense lawyers were not giving up the fight. Within view of the jurors on the porch was a giant sign labeled "Read Your Bible," which Darrow insisted on removing. Arthur Hays then stood and announced, "The defense desires to call Mr. Bryan as a witness." The prosecution lawyers immediately objected. Silent on the matter was Bryan himself.

Over the weekend, Darrow had guessed that although the prosecution was going to win the case, Bryan had not felt part of the glory. He

This 1925 photograph, taken in the Dayton courtroom during the Scopes trial, shows Clarence Darrow and William Jennings Bryan sharing a friendly conversation. Two of America's greatest public speakers, their participation on opposite sides of the Scopes case ensured the attention of the nation. Although Darrow and his team, particularly Dudley Malone, bettered the prosecution, Bryan enjoyed the advantage of support from area residents, the jury, and the judge.

had rarely spoken in court, and when he did make a speech, Dudley Malone made a better one. Over the weekend, Darrow issued a press statement to tease Bryan:

> Bryan, who blew the loud trumpet calling for a "battle to the death," had fled from the field, his forces disorganized and his pretensions exposed.

Bryan shot back at Darrow through the press. He had taken Darrow's bait.

As lawyers from both sides argued, the Great Commoner rose to his full height. He was willing to take the stand, he said, if he could cross-examine the defense attorneys. Darrow agreed. The news of the "duel to the death" ran like wildfire up and down Main Street. When Bryan sat in the witness stand, he looked out at a crowd of more than three thousand people.

HOW LONG IS A DAY?

For some time, Darrow had been waiting for this opportunity to question the leading orator of the Fundamentalist movement. Two years before, Darrow had published fifty-five questions in the *Chicago Tribune* that he wanted to ask Bryan about the Bible. These questions were the basis for the cross-examination that he was to deliver.

Darrow began slowly, "You have given considerable study to the Bible, haven't you, Mr. Bryan?"

"Yes, I have," Bryan said. "I have studied the Bible for about fifty years."

"Do you claim that everything in the Bible should be literally interpreted?"

"I believe everything in the Bible should be accepted as it is given there." Bryan then clarified that some of the language was figurative. He did not think that man was "the salt of the earth." Darrow probed him on more questions. Bryan believed that the Bible story of Jonah getting swallowed by the whale could be true because God could make it so. Darrow asked questions about science, to which the Bible and Bryan provided vague answers. Bryan was getting irritated and plainly angry when the audience laughed at his testimony.

Darrow finally got Bryan to agree to the moment when God created Earth: October 23, 4004 BC, at 9:00 AM. A biblical scholar had

Perhaps the climax of the Scopes trial, and most definitely the undoing of William Jennings Bryan, occurred when Clarence Darrow called Bryan to the stand to testify as a biblical expert. In this 1925 photo, Darrow *(standing)* examines Bryan *(seated, left, with fan)* on the porch of the courthouse. Judge Raulston moved the trial outside in part because of the stifling heat of Tennessee in July and because he was concerned the weight of the spectators would cause the courtroom floor to cave in.

calculated this date, and it was widely accepted by Fundamentalists. Darrow then inquired about the dates of some of the great civilizations. Did Bryan know that the Chinese civilization was older than that? Did he know that the Egyptians were older still?

"No," Bryan said. Nor did he know the age of the other great religions, for the Christian one satisfied him.

Darrow continued to hammer on Bryan to demonstrate his ignorance of many fields of scientific inquiry. Bryan knew nothing of archaeology, astronomy, or history. Over and over, the court, the judge, and the audience heard about how little this Fundamentalist knew. Yet, as spokesman for the Fundamentalist movement, this man claimed to know what was good for teachers to teach and for students to learn. With his questions, Darrow seemed to be asking over and over the same one to the people of Tennessee: Is this kind of ignorance what you want your schools to produce? For Bryan, it was humiliating.

Darrow returned to the Fundamentalist story of creation. He asked, "Do you think the world was created in six days?"

"Not six days of twenty-four hours," Bryan replied. A gasp went through the crowd. Here was a Fundamentalist, a believer in the literal truth of the Bible, saying that the plain text of the Bible may not be the truth. Prosecutor Stewart objected to the line of questioning, and another lengthy argument raged on the porch. Bryan defended his position: "I want the Christian world to know that any agnostic . . . can question me as to my belief in God, and I will answer him!"

This photograph captures Judge John T. Raulston reading instructions to the jury at the close of the Scopes trial in July 1925. The judge enjoyed the publicity surrounding the trial: He was always available for photo opportunities, and he offered to hold the trial in a tent in order to accommodate as many spectators as possible.

But the damage had been done. Darrow continued to question him aggressively, and the audience was now laughing at the hapless orator. Exhausted, confused, and ashamed, Bryan stepped off the witness stand. As the crowd surged forward to congratulate Darrow, Bryan slipped away quietly. The Great Commoner had lost his audience.

THE VERDICT

The next day, newspapers all across the country carried stories of the great battle on the courthouse steps and of Bryan's failure. In Dayton, Judge Raulston forbade the defense from continuing its questioning of William Jennings Bryan. Additionally, Bryan's testimony was stricken from the record of the case. Without testimony from Bryan or the expert witnesses, the defense had nothing to show that evolution was an accepted science and that outlawing the teaching of it was unconstitutional. The only hope for Scopes was to secure an appeal in a more favorable court. To do so, Darrow needed a guilty verdict. In his closing statement, he told the jury to find Scopes guilty.

The jury didn't disappoint him. Scopes was found guilty and fined $100. In his single statement during the trial, Scopes stood and said, "I feel I have been convicted of violating an unjust statute. I will continue in the future, as I have in the past, to oppose this law in any way I can." Defense attorney Hays formally requested an appeal, which Judge Raulston granted. In his final statement, Judge Raulston noted the courage of Scopes in standing for his beliefs, and the Trial of the Century ended.

CHAPTER 8

THE MEANING OF THE SCOPES TRIAL

With the end of the trial, the storm had passed through Dayton. Left behind forever was William Jennings Bryan. Five days after the trial, Bryan lay down, as he liked to do, for a nap after a typically large lunch. He died in his sleep on July 26, 1925. Some claimed that his humiliation on the witness stand had done him in, but Darrow cruelly replied, "Broken heart nothing; he died of a busted belly." William Jennings Bryan did much to give voice to those who had none. In the generations since his death, his speeches have inspired trade unions, political movements, and colleges—one college in particular. Today, Bryan College overlooks a bend in the Tennessee River in Dayton.

THE APPEAL AND THE FATE OF JOHN T. SCOPES

What had started as a summertime lark changed the life of John Scopes. Scopes was forced to stand trial again in appeal before the Tennessee Supreme Court in 1927. During his first trial, Judge Raulston had set the fine for $100, yet Tennessee law said that the

The Trial of John T. Scopes

jury must set the fine. As a result, the Tennessee Supreme Court declared that the decision in Judge Raulston's courtroom was null and void. The decision was overruled, and the Butler Act would remain a law in Tennessee for another forty-two years.

Several scientists arranged for Scopes to study geology at the University of Chicago. He left Tennessee and never taught again. He died in 1970.

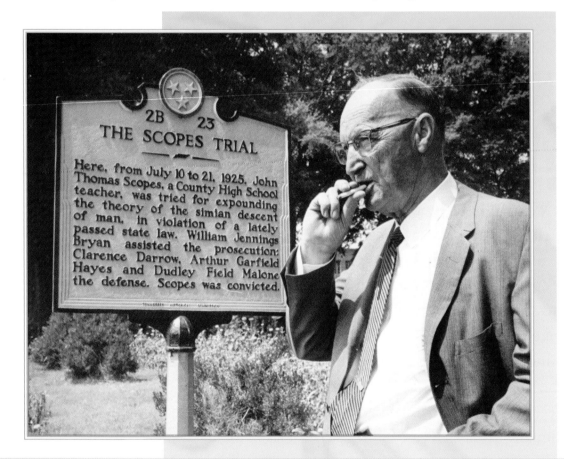

Thirty-five years after he stood trial for teaching the theory of evolution, John T. Scopes posed by a plaque marking the site in Dayton, Tennessee, that made him famous, in this news photo taken on July 21, 1960. In his memoir, *Center of the Storm* (1967), Scopes wrote, "To be sure everyone connected with the defense was positive that the U.S. Supreme Court, before whom we hoped the case would ultimately go, would render a decision in favor of the defense. How could the court do otherwise when the law infringed on academic freedom and the freedom of speech?"

After the trial, Clarence Darrow continued to practice law in Chicago. In 1932, he wrote his autobiography, *The Story of My Life*. He passed away in 1938. Clarence Darrow was a lawyer, pure and simple. While many people hated him for his frank opinions, Darrow loved the law and used it as a tool to protect the rights of the individual. Because of his willingness to represent unpopular defendants such as Eugene Debs, Nathan Leopold, and Richard Loeb, Darrow did much to test and to define the role of the individual in modern American society.

DARWIN DISAPPEARS AND THEN RETURNS

For Charles Darwin's ideas, the following years were not so promising. In Mississippi, a law similar to the Butler Act was passed in 1926. For educators, the name of Darwin inspired worry. George Hunter rewrote *A Civic Biology* to remove any references to evolution. Across the country, students interested in *The Origin of Species* had to request it at the library desk, as it was rarely on the shelves. Darwin disappeared for a long time.

In 1957, the launch of the Soviet rocket *Sputnik* into space changed opinions about science in America. In 1961, President John F. Kennedy announced that America would put a man on the moon by the end of the decade. The study of science, including biology, became a priority. Darwin returned.

Laws changed as a result of Darwin's return. In 1967, Tennessee finally repealed the Butler Act. The following year, the case of an Arkansas schoolteacher who taught evolution in class was heard before the U.S. Supreme Court. The Court ruled that the Arkansas law was unconstitutional and was so similar to the repealed Butler Act that it was repealed, too. In its ruling, the Supreme Court declared that such laws violated the First Amendment, the freedom of speech. Forty-three years after the Scopes trial, Clarence Darrow's argument finally won.

The Fundamentalists were not silenced, however. In the 1960s, Fundamentalists argued that the biblical version of creation could be supported with scientific evidence. They wanted this new field of study, called creation science, to be given equal time in the classroom. In Louisiana, a law was passed requiring teachers who taught evolution to also teach creation science. On March 19, 1981, a similar act became law in Arkansas. Many Arkansas teachers, parents, and clergymen protested that the Bible was not part of the classroom. In 1982, a court in Arkansas repealed the act, and the Arkansas Supreme Court refused to hear the appeal. Five years later, the U.S. Supreme Court declared that creation science was not science at all.

The fight between evolution and creationism, between Darrow and Bryan, continues to this day. Throughout the 1990s and into the twenty-first century, creationists sponsored legislation in multiple states in the South and the central states; all failed to become law. In 1996, presidential candidate Pat Buchanan spoke in favor of parents' rights to suppress "Godless evolution" in the classroom. On the other side stood the champions of freedom of speech. As Clarence Darrow said, "You can protect your liberties in this world only by protecting the other man's freedom. You can be free only if I am free."

In the middle of this fight stands Charles Darwin, who believed that religion and science could coexist. While science continues to make discoveries that inspire change in our lives, there will always be a need for moral guidance each and every day. Whether the story of creation in the Bible or in the scientific textbooks is true, there is a place for both in the public arena. The debate continues.

GLOSSARY

agnostic A person who believes it is impossible to know whether there is a god.

appeal A request to retry a case in a higher court.

bigot A person who is intolerant of those who differ from his or her own religion, race, gender, or political group.

civil liberties Fundamental individual rights, such as freedom of speech and religion, protected by law.

constitutional Permissible under the U.S. Constitution.

contempt of court Open disrespect and disobedience for the rules of the court.

counsel An attorney.

creationism Belief in the literal interpretation of the account of the creation of the universe and of all living things related in the Bible.

cross-examine To question a witness already examined by the opposing side.

defendant A person charged with a crime.

defense The attorney or attorneys who represent the defendant.

evolution A change in the genetic composition of a population during successive generations, resulting in the development of new species.

Fundamentalism The belief that the contents of the Bible are literally true.

grand jury A jury that evaluates accusations against persons charged with crime and determines whether the evidence warrants a bill of indictment.

indict To charge with a crime.

natural selection The process in nature by which, according to Darwin's theory of evolution, only the organisms best adapted to their environment tend to survive and transmit their genetic characteristics to succeeding generations while those less well adapted tend to be eliminated.

orator A skilled public speaker.

prosecution A lawyer or lawyers who work on behalf of the state to convict people charged with a crime.

repeal To declare that a law is no longer valid.

tabernacle A place of worship.

testimony A declaration in a criminal trial by a witness under oath.

unconstitutional A law that does not agree with the principles of the U.S. Constitution.

verdict The decision of whether a defendant is guilty or not guilty.

FOR MORE INFORMATION

The American Civil Liberties Union (ACLU)
125 Broad Street, 18th Floor
New York, NY 10004
(212) 549-2500
Web site: http://www.aclu.org

The Rhea County Courthouse & Museum
1475 Market Street
Dayton, TN 37321
(423) 775-7801

WEB SITES

Due to the changing nature of Internet links, the Rosen Publishing Group, Inc., has developed an online list of Web sites related to the subject of this book. This site is updated regularly. Please use this link to access the list:

http://www.rosenlinks.com/gttc/tjts

FOR FURTHER READING

Allen, Robert A. *William Jennings Bryan: Golden-Tongued Orator.* Fenton, MI: Mott Media, 1992.

Blake, Arthur. *The Scopes Trial: Defending the Right to Teach.* Brookfield, CT: The Millbrook Press, 1994.

Gurko, Miriam. *Clarence Darrow.* New York: HarperCollins Children's Books, 1965.

Hanson, Freya Ottem. *The Scopes Monkey Trial.* Berkeley Heights, NJ: Enslow Publishers, 2000.

McGowen, Tom. *The Great Monkey Trial.* New York: Franklin Watts, 1990.

Nardo, Don. *The Scopes Trial.* San Diego, CA: Lucent Press, 1997.

For Advanced Readers

Darrow, Clarence. *The Story of My Life.* New York: Charles Scribner's Sons, 1932.

Weinberg, Arthur (ed.). *Attorney for the Damned.* Chicago: University of Chicago Press, 1989.

BIBLIOGRAPHY

Blake, Arthur. *The Scopes Trial: Defending the Right to Teach*.
   Brookfield, CT: The Millbrook Press, 1994.

Chapman, Matthew. *Trials of the Monkey: An Accidental Memoir*. New
   York: Picardor USA, 2000.

Darrow, Clarence. *The Story of My Life*. New York: Charles Scribner's
   Sons, 1932.

Grant, Robert, and Joseph Katz. *The Great Trials of the Twenties*.
   Rockville Center, NY: Sarpedon Press, 1998.

Hanson, Freya Ottem. *The Scopes Monkey Trial*. Berkeley Heights, NJ:
   Enslow Publishers, 2000.

Larson, Edward J. *Summer for the Gods: The Scopes Trial and
   America's Continuing Debate Over Science and Religion*. New York:
   BasicBooks, 1997.

McGowen, Tom. *The Great Monkey Trial*. New York: Franklin
   Watts, 1990.

Nardo, Don. *The Scopes Trial*. San Diego, CA: Lucent Press, 1997.

Public Broadcasting Service. "Monkey Trial," *American Experience*.
   1999. Retrieved June 29, 2002 (http://www.pbs.org/wgbh/
   amex/monkeytrial).

PRIMARY SOURCE IMAGE LIST

Cover and page 45: News photograph of Clarence Darrow and William Jennings Bryan in Dayton, Tennessee, taken in 1925. Housed in the Bryan College Archives.

Page 4: Photographic portrait of John T. Scopes, 1925. Housed in the Bryan College Archives.

Page 6: Photograph of the reenactment of the meeting of the Scopes conspirators in Robinson's Drugstore in Dayton, Tennessee, taken on July 6, 1925.

Page 8: Photograph of John Washington Butler, taken at his desk in the Tennessee House of Representatives. Housed in the Bryan College Archives.

Page 9: Cover and interior pages (193 and 194) of George William Hunter's *A Civic Biology*. Published in 1914.

Page 12: Photograph of George Rappelyea and John Scopes in Dayton, Tennessee, taken on July 18, 1925.

Page 14: Photographic portrait of Charles Darwin. Housed in the Library of Congress. Photograph of *On the Origin of Species* (1859), by Charles Darwin. Drawing of a gorilla, Henry Bergh, and Charles Darwin, by Thomas Nast, circa 1871.

Page 19: Photograph of William Jennings Bryan at the pulpit of the Little Methodist Church in Dayton, Tennessee, taken in July 1925.

Page 20: Photograph of William Jennings Bryan meeting with men during the Scopes trial in South Dayton, Tennessee, on June 23, 1925.

Page 21: Photographic portrait of Clarence Darrow, taken in 1922. Housed in the Library of Congress.

Page 22: Photograph of John Scopes with Clarence Darrow and legal staff. Taken in 1925 in Dayton, Tennessee.

Page 25: Photograph of crowds flocking to street vendors in Dayton, Tennessee, taken on July 7, 1925. Housed in the Bryan College Archives.

Page 26: Photograph of Lena Ruffner holding a monkey doll. Taken on July 11, 1925, in Dayton, Tennessee.

Page 28: Photographic portrait of Judge John Raulston, taken on July 11, 1925, in Dayton, Tennessee. Housed in the Bryan College Archives.

Page 29: Photograph of Scopes jury members, taken in 1925 in Dayton, Tennessee.

Page 32: News photograph of William Jennings Bryan and others in the courtroom praying. Taken on July 15, 1925, in Dayton, Tennessee.

Page 34: Photograph of antievolution league book stand, taken in the summer of 1925 in Dayton, Tennessee.

Page 35: Photograph of Walter White testifying, taken on July 15, 1925, in Dayton, Tennessee.

Page 39: Photograph of Scopes trial notables, taken in July 1925 in Dayton, Tennessee. Photograph of Maynard Metcalf taken in 1925 in Dayton, Tennessee.

Page 42: Photograph of William Jennings Bryan during the Scopes trial. Taken in 1925 in Dayton, Tennessee. Drawing of William Jennings Bryan shining down on the town of Dayton, drawn by Clifford Kennedy Berryman in 1925.

Page 47: Photograph of Clarence Darrow questioning William Jennings Bryan on the courthouse porch in Dayton, Tennessee. Taken in July 1925.

Page 49: Photograph of Judge John Raulston, taken in 1925 in Dayton, Tennessee.

Page 52: Photograph of John T. Scopes in Dayton, Tennessee, taken on July 21, 1960.

INDEX

ABOUT THE AUTHOR

Steven P. Olson is a freelance writer who lives in Oakland, California, and likes to travel the world. His Web site can be found at http://www.stevenolson.com.

CREDITS

Cover, pp. 1, 6, 20, 32, 45 © Underwood & Underwood/Corbis; pp. 4, 8, 9, 25 Courtesy of Bryan College Archives; pp. 12, 19, 28, 34 © Hulton/Archive/Getty Images; p. 14 (top left) Library of Congress Rare Book and Special Collections Division; pp. 14 (top right), 21, 42 (inset) Library of Congress Prints and Photographs Division; p. 14 (bottom) © Corbis; pp. 22, 26, 29, 35, 39, 42 (left), 47, 49, 52 © Bettmann/Corbis.

The July 23, 1925, front page of the *Dayton Herald* on the cover is reprinted with permission of the *Herald-News*, Dayton, Tennessee.

ACKNOWLEDGMENTS

The publisher would like to thank Tom Davis and Bryan College for their help with some of the photos in this book.

Designer: Les Kanturek; **Editor:** Christine Poolos.